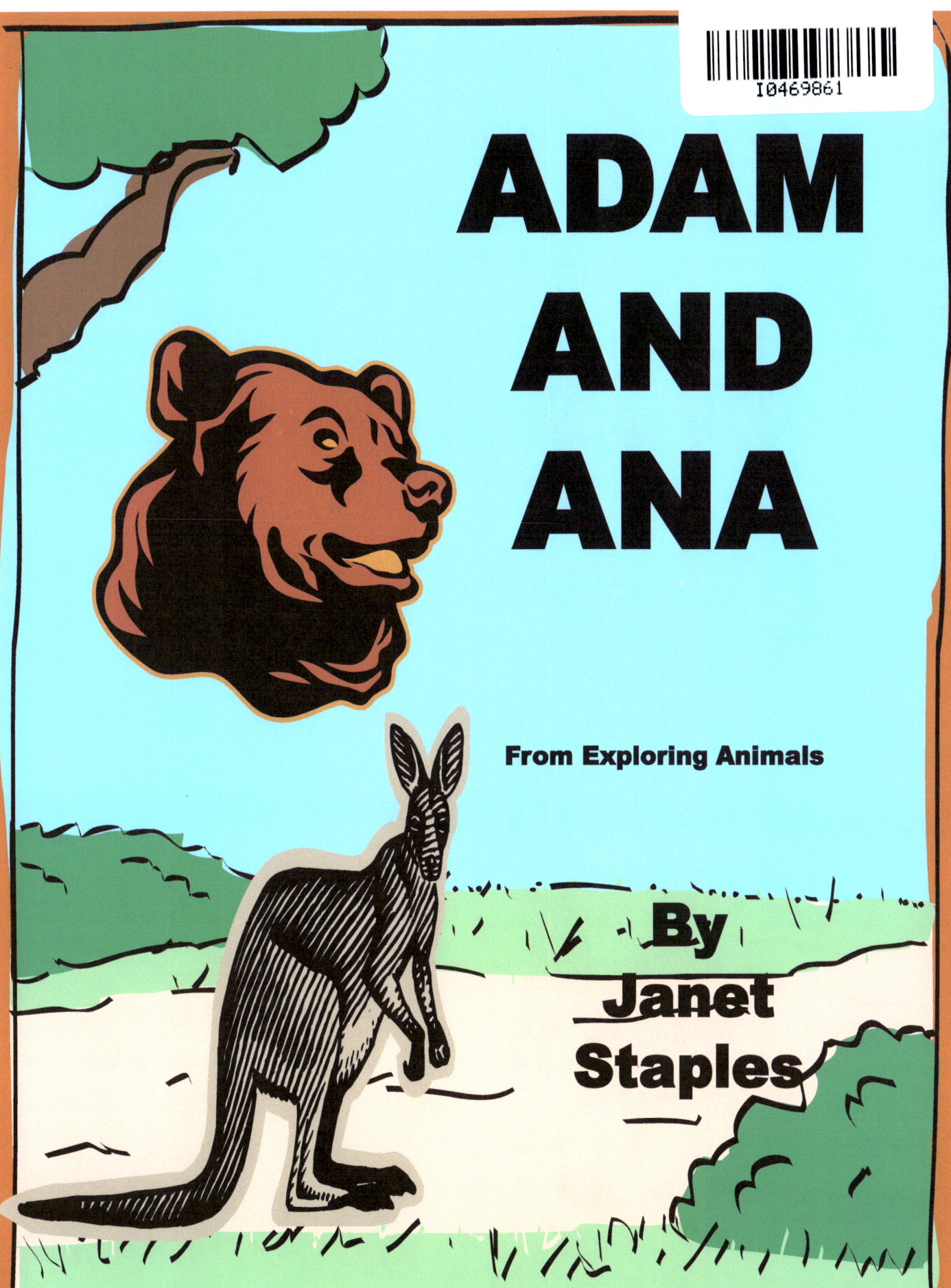

ADAM AND ANA

From Exploring Animals

By
Janet Staples

TABLE OF CONTENTS

About the Author

- _**Janet Staples:**_ Author, educator, specialist in multi-sensory-multi-disciplinary learning. With thirty-eight years of classroom experience, her innovative and effective techniques have redefined success for thousands of Virginia's elementary and middle school students.

- _**A Few of My Presentations:**_Tele Conference for Math tea

- chers at Old Dominion University.

- Teaching Strategies to Math and Science teachers: ODU, 2002

- 7th, 8th, 9th, and 10th Old Dominion University Annual Symposium.

- 9th Annual Mid-Atlantic Symposium on SOLS: Old Dominion University 2001.

- 333 County Schools 2001.

- Dan River Elementary and Middle School: Danville, Virginia 2001.

- Commonwealth of Virginia Department of Correctional Education: Richmond, Virginia, 2000.

- Regional Math Conference, New Jersey, 2001.

- Tidewater Math Conference: Maury High School, 2001.
  - Peninsula Math Conference: 2001

- Pittsylvania County Schools: Danville, Virginia, 2000

- Mathews Public Schools.

- _**Here is what they have been saying:**_

- I appreciate you taking the time to share with teachers your ideas and activities. I heard nothing but good comments from teachers as they moved from session to session. Everyone seemed excited about learning new ways to help students learn. Elizabeth M. O'Brien: Math Coordinator, Peninsula Council.

- Best in service in 30 years of teaching. Participant Virginia Commonwealth for Correctional Education.

- Invite her back. Super session. We need more teachers like Janet. Participant at the Mid--Atlantic Symposium on SOLS.

- Activity Puts Kids On Hunt For Knowledge: Margaret Windley: Portsmouth Times.

- Staples has combined her idea for the Treasure Hunt and other games in books that she uses in seminars throughout Hampton Roads: Stephanie Crockett: Virginia Pilot.

- On behalf of the Instructional Team, may I take this opportunity to say thank you for assisting Pittysylvania County Schools at the Summer Institute. J. E. Daniels: Administrative Assistant Instruction, Pittyslvania County Schools.

- Teacher Focus On Fun And Sols: Tina Mccloud: The Daily Press.

- ## I would like to thank my family and those educators that supported my efforts.

- ## Janet L. Staples: State of Virginia Educator for 38 years.

Student Outcomes

- Educational Outcomes for Students K-3:
- *SCIENCE*
- The student will read and comprehend that animals have:
- Specific physical characteristics.
- Specific adaptations.
- Specific behaviors.
- Specific needs.
- Specific interaction with other animals.
- Specific habitats.
- Off springs.
  - *Geography:* Student will identify states by shapes and locations on a map.
- *Geography:* Student will identify continents.
- *History*: Student will learn that states have their own resources.
- *History:* Student will identify state symbols.
- *History:* Student will learn facts about states.
- *Reading:* Student will describe people, places, and things.
- *Reading:* Student will be able to retell the story.
- *Reading:* Student will increase their vocabulary.
- *Reading:* Student will identify pictures as clues to help identify words.

Adam

Australia

My name is Adam and I live in Australia.

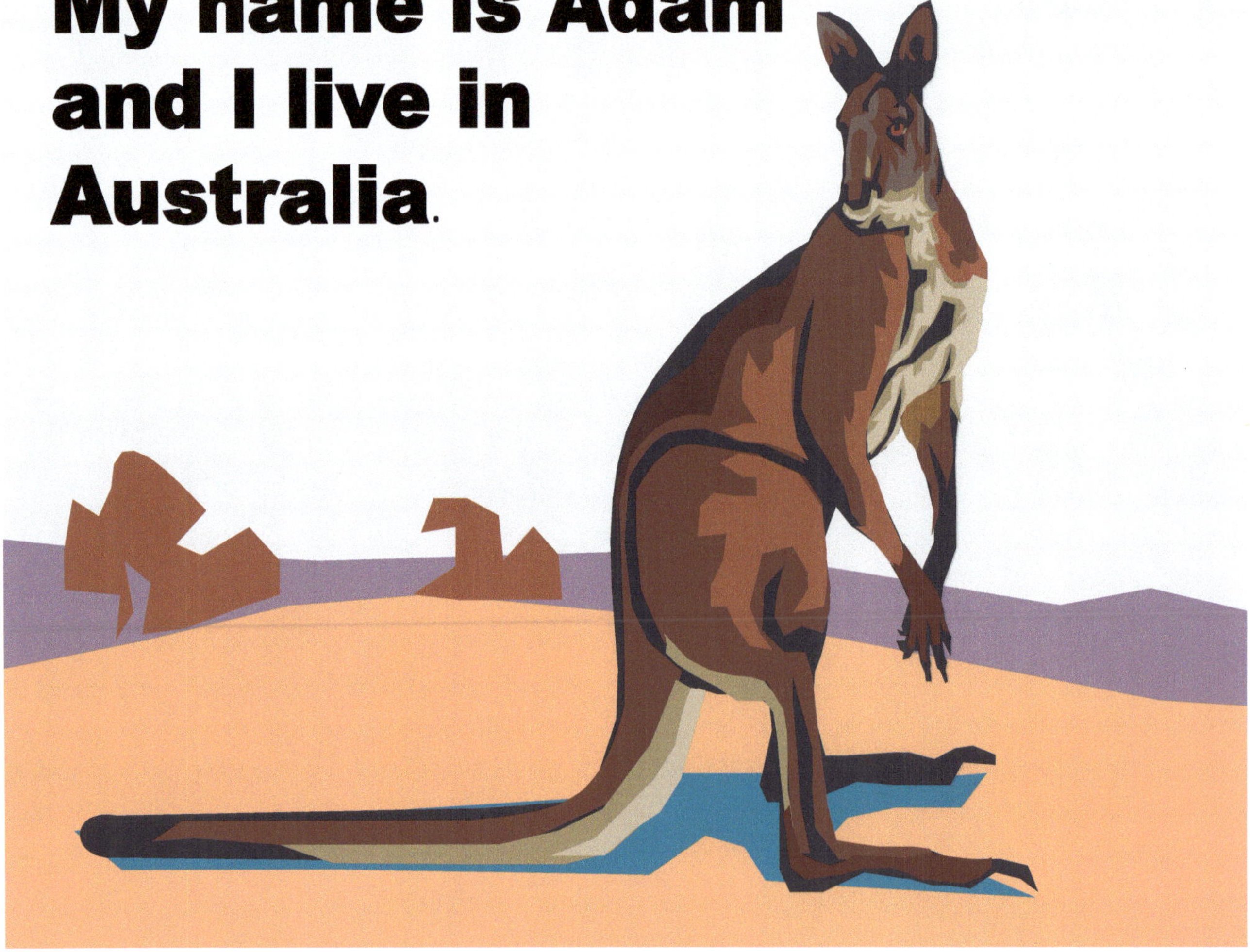

Australia is a continent, surrounded by the Indian and Pacific Ocean.

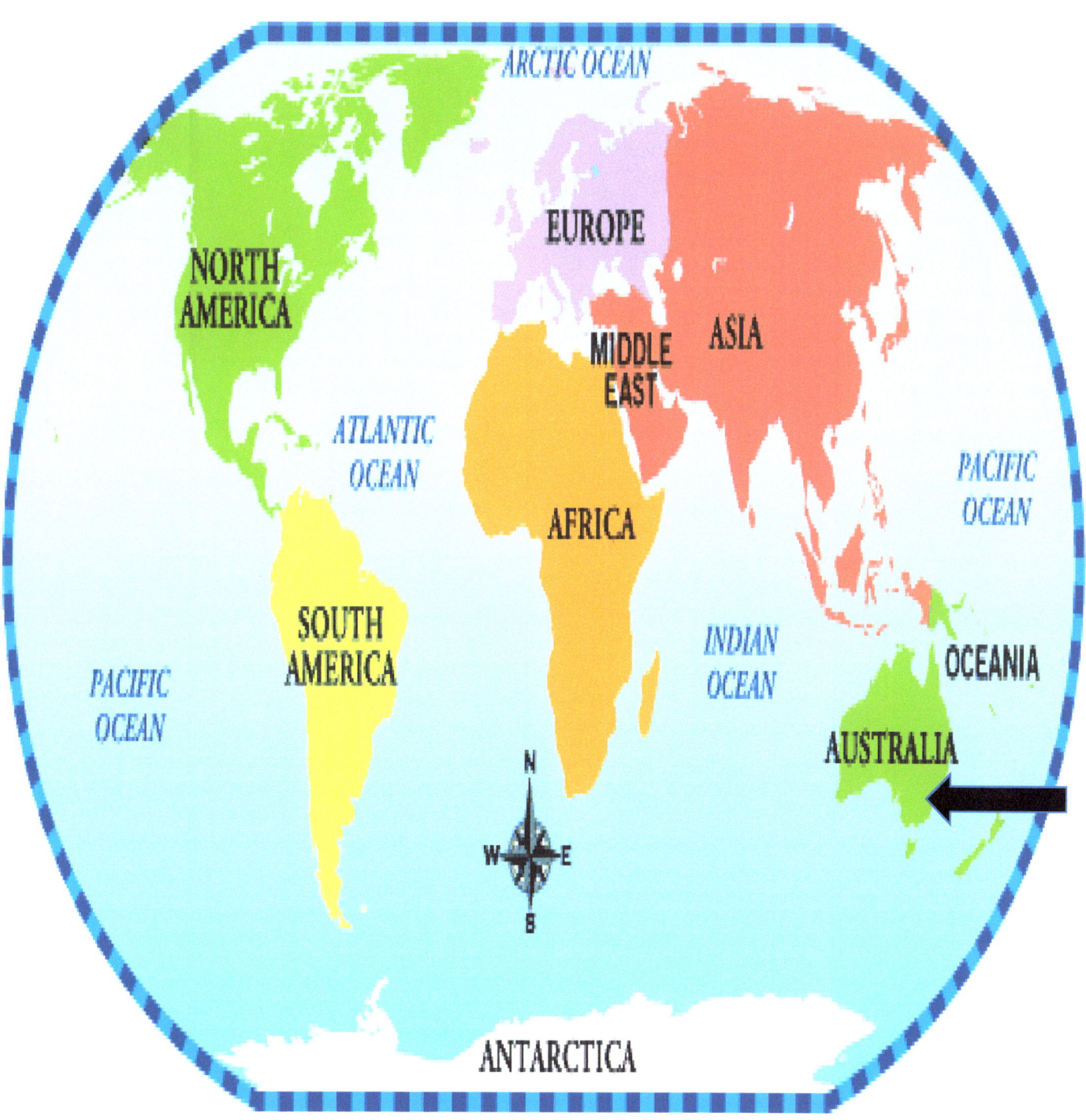

I am a male kangaroo. They call me Jack, Boomer or Buck.

I have a large tail. I can not walk so I use my tail to help me jump. I can jump up to 30 feet.

We are Marsupials, because the females have a pouch. The females are called, Jills, Does or Flyers.

I live in small groups of other kangaroos, called mobs.

We are grazing herbivores, which means our diet consists mainly of grasses.

Every year the females have babies. The mothers have a pouch where she carries the babies. The babies are called Joeys. They stay in her pouch for about 4 months.

I will live for about 14 years. It is very hot in Australia and the heat causes drought and hunger.

ANA

Hello, My name is Ana and I am a Brown Bear. Sometimes they call me a Grizzly Bear. I live in Europe and North America.

I have been asleep for 6 months. This is called hibernation and I take this nap every year. I will wake up in the spring. This is when the birds return from their migration.

Today is my birthday and I am 8 years old. I will live to be 15 to 30 years old.

Here comes George. He lives in Yellowstone National Park. George is a big brown bear and weighs over 1,000 pounds. George has large paws with five claws.

Most brown bears have two cubs in the spring. These are our children, George Jr. and Susie. They will live with us for about two years.

George Jr. and Susie will live with us until they can take care of themselves. They will learn to fish, climb trees, and protect themselves.

George and I are proud of Susie and George Jr. We know they will be able to take care of themselves. It is time for them to leave and find homes of their own.

I will miss them, but we will keep in touch. Here comes winter and it is time for my nap.

George will go back to Yellowstone
National Park and come back next year.

Good bye. I'll see you
in the spring.

THE END